ARE YOU AFFECTED??

The AI Revolution That Takes Our Jobs and Changes the Way We Live

Vance Jefferson

Table of Contents

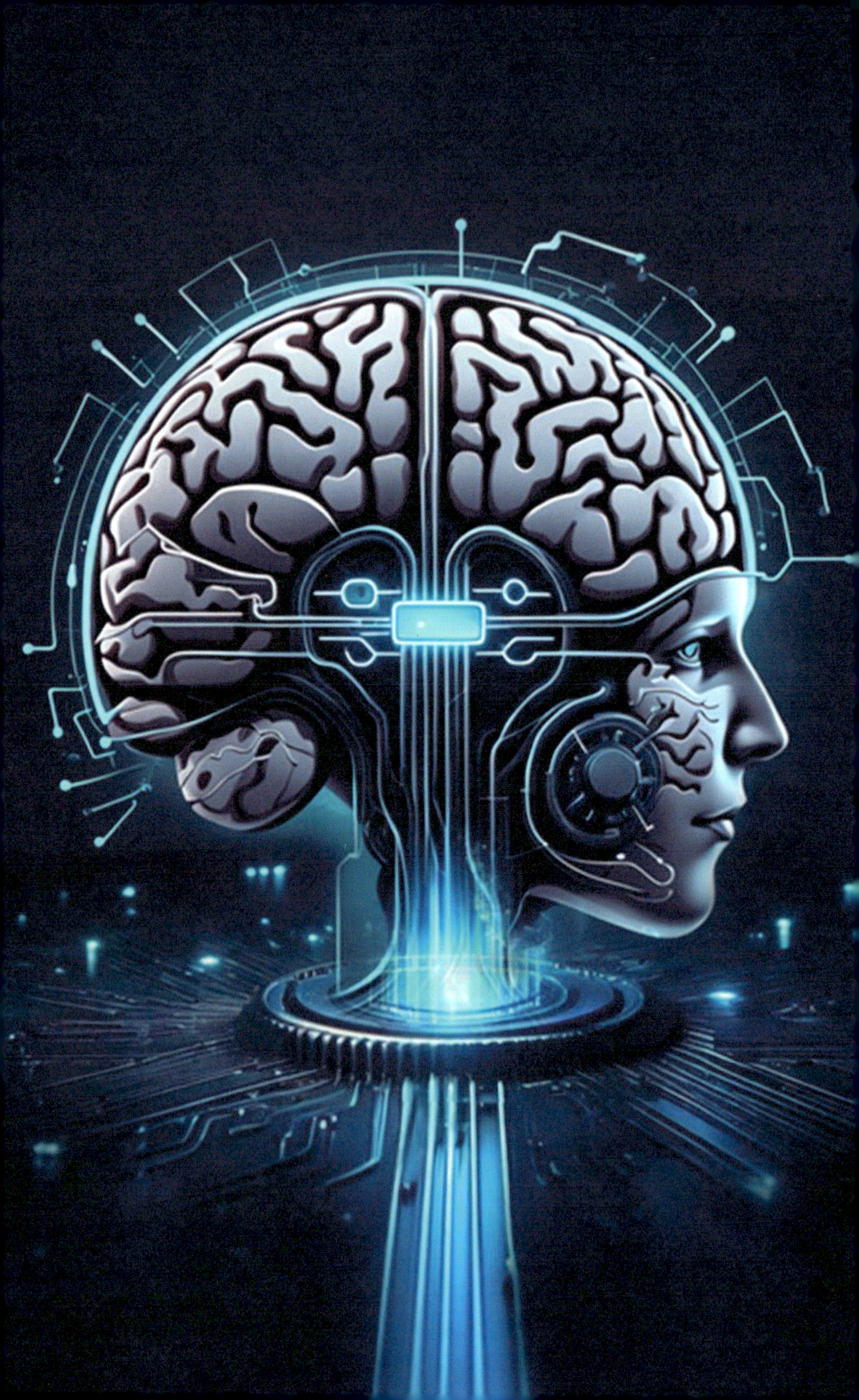

Chapter 1: A New Age of Technology Has Arrived! Is It Here to Stay?

Introduction to AI

Artificial Intelligence (AI) is transforming our world at a breathtaking pace. Imagine a world where your car drives itself to work; your virtual assistant schedules your meetings, and your refrigerator orders groceries before you run out. This is not the future—this is now. AI is here, and it's reshaping our daily lives in profound ways.

The Evolution of AI

The concept of AI dates back to the mid-20th century, but significant advancements have only occurred in recent decades. Initially, AI research focused on simple problem-solving and symbolic methods. However, AI has evolved dramatically with the advent of powerful computing resources and large datasets. Modern AI, driven by machine learning and deep learning, can handle complex decision-making and predictive analytics.

Significance of AI

AI's impact spans numerous industries, revolutionizing how businesses operate and deliver services. In healthcare, AI aids in diagnosing diseases, personalizing treatment plans, and predicting patient outcomes. In finance, AI algorithms detect fraudulent transactions and manage investment portfolios. The automotive industry is experiencing a shift with the development of autonomous vehicles, enhancing road safety and efficiency.

In our daily lives, AI manifests in various forms, such as virtual assistants like Siri and Alexa, recommendation systems on platforms like Netflix and Amazon, and even smart home devices that automate our living environments. These applications demonstrate AI's potential to improve convenience and productivity.

A Personal Story

Meet the Johnson family: Jane, a schoolteacher; Tom, a factory worker; and their two children, Emma and Jack. The Johnsons live in a typical suburban neighborhood, where AI has quietly but profoundly changed their lives. Jane uses an AI-powered platform to create personalized lesson plans for her students. Tom, who faced job insecurity due to factory automation, retrained as a technician to maintain the robots that replaced his old job. Emma and Jack use AI-driven educational tools that adapt to their learning styles, making homework a breeze.

Conclusion

The future of AI holds even more promise. AI will likely become more sophisticated, integrated, and indispensable as technology advances. The question remains: is this technological revolution here to stay? Given the ongoing investments, research, and integration of AI into critical systems, it is evident that AI is not a fleeting trend but a fundamental shift that will shape the future.

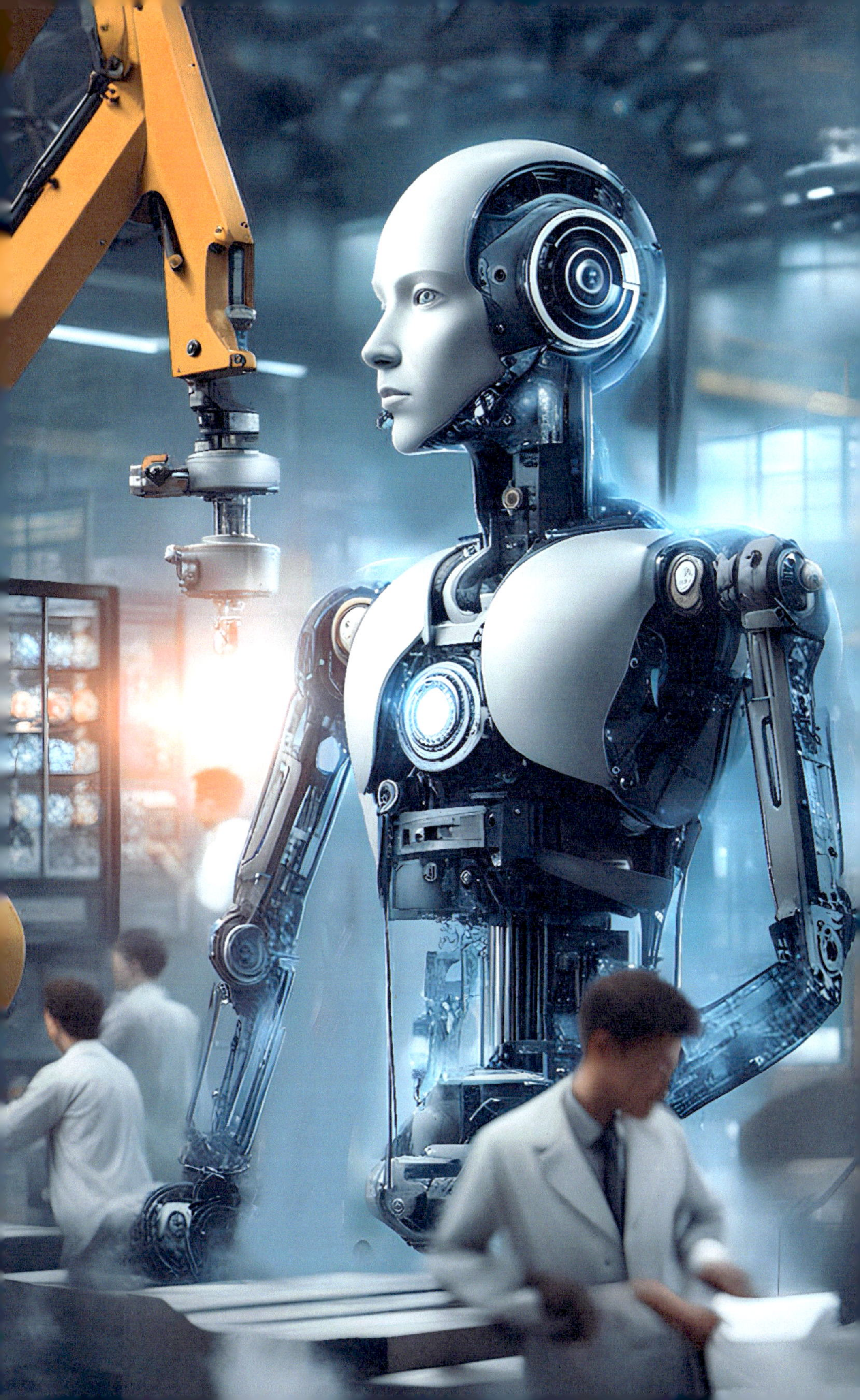

Chapter 2: What is AI, and Do We Need It, or Can We Live Without It?

Detailed Explanation of AI

AI encompasses a broad range of technologies designed to perform tasks that typically require human intelligence. These include recognizing speech, learning from data, making decisions, and translating languages. AI can be categorized into three types:

Narrow AI: Also known as Weak AI, this type is designed to perform a narrow task (e.g., facial recognition or internet searches). Most of the AI applications we interact with daily fall into this category.

General AI: Also referred to as Strong AI, this type of AI can understand, learn, and apply knowledge across various tasks, similar to a human being. General AI remains largely theoretical and is a major goal of ongoing research.

Super intelligent AI: This hypothetical form of AI surpasses human intelligence across all fields. While still in science fiction, discussions around super intelligent AI often focus on potential risks and ethical considerations.

Machine learning, a subset of AI, involves training algorithms to learn patterns from data and make decisions based on it. Neural networks, inspired by the human brain's structure, form the basis of deep learning, enabling advanced tasks such as image and speech recognition.

Simplifying AI Concepts

Think of AI as a super-efficient, tireless assistant. Narrow AI is like a specialized assistant skilled in one area—like a master chef who can cook any dish perfectly but can't manage your finances. General AI would be like an all-rounder assistant capable of handling various tasks with human-like versatility. If it existed, Super intelligent AI would be an assistant whose capabilities far exceed human intelligence.

Necessity of AI

AI offers numerous benefits that are becoming increasingly essential in today's fast-paced world. For instance, AI-powered tools enhance productivity by automating repetitive tasks, allowing humans to focus on more creative and strategic work. In healthcare, AI-driven diagnostics can lead to earlier detection of diseases and personalized treatment plans, improving patient outcomes.

AI also plays a crucial role in addressing complex global challenges. For example, AI models predict climate change patterns, helping policymakers create more effective environmental strategies. AI optimizes crop yields and resource use in agriculture, contributing to food security.

Despite these advantages, the necessity of AI can be debated. Some argue that reliance on AI may lead to job displacement and exacerbate social inequalities. However, others contend that AI's potential to drive innovation, efficiency, and solutions to global issues outweighs these concerns. Educating ourselves about AI is vital to navigating its impacts responsibly and harnessing its benefits effectively.

Practical Advice

- **Learn AI Basics:** Take online courses to understand AI fundamentals.
- **Stay Informed:** Follow AI news to stay updated on new developments.
- **Embrace AI Tools:** Use AI-driven apps and tools to enhance productivity in daily life.

Chapter 3:
The Upside and Downside of AI

Advantages of AI

AI technology brings several benefits that enhance various aspects of life and business operations:

- **Increased Efficiency and Productivity:** AI systems can process vast amounts of data quickly and accurately, making them invaluable for tasks such as data analysis, customer service, and logistics. Routine task automation frees human workers to focus on more strategic and creative endeavors, boosting overall productivity.
- **Innovations in Healthcare:** AI has revolutionized healthcare by enabling precise diagnostics, personalized treatment plans, and efficient patient management. AI algorithms analyze medical images to detect diseases early, and predictive models anticipate patient outcomes, leading to better care and treatment.
- **Enhanced Decision-Making Processes:** AI tools support decision-making by providing insights based on data analysis. Businesses use AI to forecast trends, optimize supply chains, and improve customer experiences. In finance, AI-driven models predict market movements and assess risks more accurately.

Disadvantages of AI

Despite its benefits, AI also poses significant challenges and risks:

- **Job Displacement and Economic Impacts:** One of the most pressing concerns is the potential for AI to displace jobs. Automating tasks previously performed by humans can lead to unemployment and economic disruption. Industries such as manufacturing, retail, and transportation are particularly vulnerable to job losses due to AI automation.

- **Ethical and Privacy Concerns:** AI systems often rely on large datasets, which raises privacy issues. The use of personal data for AI-driven decision-making can lead to breaches of privacy and misuse of information. Additionally, AI algorithms can perpetuate biases in training data, leading to unfair outcomes in hiring, law enforcement, and lending.
- **Dependence on Technology:** As AI becomes more integrated into daily life, there is a risk of over-reliance on technology. This dependency can lead to the loss of critical thinking skills and reduce human intervention in important decision-making processes. Furthermore, technical failures or malicious attacks on AI systems could have catastrophic consequences.

Case Study: AI in Healthcare

Consider the story of Sarah, a cancer patient who benefitted from AI-driven healthcare. Traditional diagnostic methods missed the early signs of her illness, but an AI system detected anomalies in her medical scans that led to an early diagnosis and treatment plan. This improved her prognosis and highlighted how AI can save lives.

Practical Tips

- **Adapt and Learn:** Stay updated with AI advancements in your field.
- **Engage Ethically:** Advocate for ethical AI practices and data privacy.
- **Balance Technology Use:** Use AI tools to enhance but not replace human decision-making.

Chapter 4: AI Takes Our Jobs, We Have Our Families to Feed, Our Newborn, Our Parents and Grandparents. Who Is Responsible?

Job Displacement

The rise of AI and automation has led to significant changes in the job market. Industries such as manufacturing, transportation, and retail are experiencing profound shifts as machines and algorithms take over tasks traditionally performed by humans. For example, self-checkout systems in retail stores reduce the need for cashiers, and autonomous vehicles threaten the livelihood of truck drivers and taxi operators.

Personal Story: Tom's Journey

A former factory worker, Tom faced job displacement when his plant introduced AI-driven automation. Initially, he struggled with the fear of unemployment and the pressure of supporting his family. However, Tom decided to embrace change. He enrolled in a retraining program funded by a government initiative and became an AI maintenance technician. Now, Tom supports his family and enjoys job security and a new sense of purpose.

Examples of Industries Affected by AI:

- Manufacturing: Automation of assembly lines and quality control processes.
- Transportation: Development of self-driving cars and drones.
- Retail: Introduction of AI-powered customer service and inventory management systems.

Emotional and Societal Impact

The impact of AI on employment goes beyond economic concerns; it also affects individuals and families on a deeply personal level. Job loss can lead to stress, anxiety, and uncertainty about the future. Families relying on a single income may struggle to meet ends, and the pressure to adapt to new technologies can be overwhelming.

Practical Advice

- **Seek Retraining Opportunities:** Look for programs that offer skills training in AI and related fields.
- **Support Networks:** Build a support network of family, friends, and professional counselors to navigate the emotional impact of job loss.
- **Advocate for Policies:** Engage with policymakers to promote policies that support retraining and job placement for displaced workers.

The Role of Stakeholders

Addressing the challenges of AI-induced job displacement requires collaboration among various stakeholders:

- **Governments:** Develop and fund retraining programs, provide unemployment benefits, and create policies encouraging ethical AI use.
- **Corporations:** Invest in employee upskilling and create opportunities for human-AI collaboration. Corporations should also prioritize ethical AI deployment that considers the social impact.
- **Educational Institutions:** Adapt curricula to prepare future generations for an AI-driven world. This includes integrating AI and digital literacy into education systems.

HEALTH CARE
CENTER

Chapter 5: AI Has Isolated Our Families, As Everyone Grabs Their Laptop and Smartphone and Digs Themselves into a Quiet Place, Updating Themselves with AI

Social Isolation Due to AI

The pervasive presence of AI in our lives has led to increased screen time and reduced face-to-face interactions. Family members often find themselves engrossed in their devices, consuming AI-driven content and engaging with virtual assistants. This trend can lead to isolation, even when physically present in the same space.

Story of the Smith Family

The Smith family, once known for their lively dinner conversations, is now caught in the technology web. Each member is glued to their device, whether AI-driven games for the kids, AI-based work tools for the parents, or social media recommendations tailored by AI. The sense of togetherness has dwindled, replaced by quiet isolation.

Examples of Families Experiencing Isolation:

- A household where each member is absorbed in their digital devices during dinner time, leading to minimal conversation and bonding.
- Parents and children prefer online activities over shared family experiences, weakening familial connections.

Importance of AI Familiarity

Understanding and using AI responsibly can bring families closer together despite these challenges. Familiarity with AI can enhance family life by facilitating better communication, entertainment, and productivity.

Benefits of Understanding and Using AI:

- AI-powered tools can help manage household tasks, freeing time for family activities.
- Educational AI applications can support children's learning and development, allowing parents to participate in their education.

Practical Tips

- **Set Tech-Free Times:** Designate times when no one uses digital devices, such as during meals or family outings.
- **Collaborate with AI:** Use AI applications that promote family activities, like cooking together using AI recipe suggestions or playing AI-driven educational games.
- **Balance AI Use:** Encourage a balance between digital and real-world interactions to maintain strong family bonds.

Chapter 6:
The AI Gold rush. Will Every Thinking Person Become a Millionaire?

Opportunities in AI

The AI revolution has sparked what some call a "gold rush," with numerous opportunities for individuals to achieve financial success through mastery of AI technologies. The demand for AI skills is skyrocketing, creating lucrative career paths and entrepreneurial ventures.

Profitable Career Paths in AI:

- **Data Scientists and Machine Learning Engineers:** These professionals design and implement AI models, earning substantial salaries due to the high demand for their expertise.
- **AI Consultants:** Specialists who advise businesses on integrating AI into their operations can command high fees for their services.
- **AI Entrepreneurs:** Innovators who develop AI-based products or services can achieve significant financial success. For instance, startups focused on AI-driven healthcare solutions or automation tools are attracting massive investments.

Challenges and Considerations

While the potential for financial gain is significant, it is essential to approach the AI gold rush with realistic expectations and a clear understanding of the challenges involved.

Skills and Knowledge Required for Success:

- Mastering AI technologies requires a strong foundation in mathematics, computer science, and domain-specific knowledge.
- Continuous learning and staying updated with the latest advancements in AI are crucial for maintaining a competitive edge.

Realistic Expectations and Potential Risks:

- Not everyone will become a millionaire; success in AI, like any field, requires dedication, hard work, and sometimes a bit of luck.
- The AI market is competitive, and failure is a possibility. Entrepreneurs must be prepared to navigate setbacks and adapt to changing market conditions.

Practical Advice

- **Learn AI Skills:** Invest in education and training to acquire in-demand AI skills.
- **Network with Experts:** Join AI communities and attend conferences to stay informed and build professional connections.
- **Stay Adaptable:** Be prepared to pivot and adapt your strategies as the AI landscape evolves.

Chapter 7: What's the Future of AI?

Future AI Innovations

We can expect significant advancements and new applications across various fields as AI technology evolves.

Expected Advancements in AI Technology:

- **Improved Natural Language Processing (NLP):** Future AI systems will understand and generate human language more accurately, making machine interactions more intuitive and seamless.
- **Enhanced Computer Vision:** AI will achieve higher precision in interpreting visual data, leading to advancements in autonomous vehicles, medical imaging, and security.
- **General AI Development:** Research efforts aim to create AI systems that possess general intelligence and are capable of performing a wide range of tasks with human-like understanding and flexibility.

Potential New Applications of AI:

- **AI in Education:** Personalized learning experiences tailored to individual student needs and preferences.
- **Environmental AI:** Technologies to monitor and mitigate environmental issues, such as climate change and wildlife conservation.
- **Smart Cities:** AI-driven infrastructure that enhances urban living through efficient energy management, traffic control, and public safety.

Societal Impact

The long-term effects of AI will reshape various industries and the job market.

Long-Term Effects on Various Industries:

- **Healthcare:** Continued improvements in diagnostics, treatment, and patient care, leading to better health outcomes and reduced healthcare costs.
- **Finance:** More sophisticated financial models and risk management tools, enhancing economic stability and growth.
- **Manufacturing:** Increased automation and precision, resulting in higher productivity and cost efficiency.

Predictions for the Future Job Market:

- AI will create new job opportunities in AI ethics, AI safety, and AI-human collaboration.
- Workforce transformation will require reskilling and upskilling programs to prepare individuals for AI-integrated roles.

Expert Opinion

Dr. Susan Lee, an AI researcher, predicts that "AI will become an integral part of our daily lives, much like electricity. It will drive innovation in ways we can't fully imagine, but we must address the ethical and social implications head-on."

Practical Tips

- **Stay Updated:** Follow AI developments to understand their potential impact on your industry.
- **Engage in Discussions:** Participate in forums and discussions about AI's future to stay informed and contribute to the dialogue.
- **Prepare for Change:** Proactively acquire new skills and adapt to the evolving job market.

Chapter 8: Should We Fight Against AI or Is It Inevitable?

Resistance to AI

As AI technology advances, some argue that we should resist its development due to potential negative consequences.

Arguments for Fighting AI Development:

- **Job Loss:** Concerns about widespread unemployment as AI replaces human workers in various industries.
- **Privacy and Security:** Fears that AI could lead to increased surveillance and data breaches.
- **Ethical Dilemmas:** AI decision-making, accountability, and potential biases.

Embracing AI

Despite these concerns, many believe that embracing AI is inevitable and beneficial.

Benefits of Accepting and Adapting to AI:

- **Economic Growth:** AI-driven innovation can boost productivity and create new markets.
- **Improved Quality of Life:** AI applications in healthcare, education, and daily life can enhance well-being and convenience.
- **Solving Complex Problems:** AI has the potential to address global challenges such as climate change, poverty, and disease.

Balancing AI's Impact

Whether we fight against AI or embrace it, proactive measures are necessary to navigate its impact on society. We can ensure that AI contributes positively to our future by addressing the challenges and maximizing the benefits.

Practical Advice

- **Advocate for Ethical AI:** Support policies and initiatives that promote ethical AI development and usage.
- **Educate Yourself:** Stay informed about AI advancements and their potential impact on society.
- **Be Adaptable:** Embrace change and adapt to new technologies and opportunities.

Chapter 9: Should We Argue That If AI Takes Away 2 Billion Jobs, But Turns 4 Billion People into Millionaires, Are Justifiable?

Economic Disruption and Wealth Creation

The rapid adoption of AI technology has the potential to disrupt traditional economic structures, leading to significant job losses while simultaneously creating new opportunities for wealth generation.

Job Losses Due to AI:

- Automation of routine tasks and processes can lead to significant job displacement in industries such as manufacturing, retail, and transportation.
- AI-driven technologies may replace roles that require repetitive or easily modifiable tasks.

Wealth Creation Through AI:

- New industries and markets driven by AI innovation can create substantial wealth for entrepreneurs, investors, and skilled professionals.
- AI can enhance productivity and efficiency, leading to economic growth and increased profits for businesses that effectively leverage the technology.

Ethical and Social Considerations

While the economic potential of AI is immense, the ethical and social implications of widespread job displacement must be carefully considered.

Ethical Questions:

- Is it morally acceptable to prioritize wealth creation over job preservation? How can we balance economic growth with social responsibility?
- How should the benefits of AI-driven wealth be distributed to ensure that all members of society can participate in and benefit from economic gains?

Strategies for Mitigating Negative Impacts

- **Social Safety Nets:** Implementing policies such as universal basic income or unemployment benefits to support those displaced by AI.
- **Reskilling and Education:** Investing in education and training programs to help workers transition to new roles in the AI-driven economy.
- **Inclusive Growth:** Ensuring that the economic benefits of AI are widely distributed, reducing income inequality and promoting social cohesion.

Practical Advice

- **Advocate for Fair Policies:** Support policies that ensure fair distribution of AI's economic benefits.
- **Stay Informed:** Keep up with AI trends to understand potential economic impacts.
- **Engage in Dialogue:** Participate in discussions about the ethical implications of AI to contribute to balanced decision-making.

Chapter 10: What Will Our Future Be Like with This AI Boom?

Transformative Effects of AI

The widespread adoption of AI technology will have profound and far-reaching effects on various aspects of our lives.

Healthcare Transformation:

- AI-driven diagnostics and personalized treatment plans will improve patient outcomes and more efficient healthcare delivery.
- Telemedicine and remote monitoring will increase access to healthcare, particularly in underserved areas.

Education and Learning:

- Personalized learning experiences tailored to individual student needs and preferences will enhance educational outcomes.
- AI-driven tools and platforms will provide people worldwide access to high-quality education, regardless of geographic location.

Daily Life and Convenience:

- AI-powered smart homes will automate routine tasks, enhancing convenience and improving quality of life.
- Autonomous vehicles and AI-driven transportation systems will make travel safer and more efficient.

Preparing for an AI-Driven Future

To fully realize the benefits of the AI boom, individuals, businesses, and governments must proactively prepare for the changes it will bring.

Skills and Education:

- Emphasizing STEM (science, technology, engineering, and mathematics) education to prepare the workforce for AI-driven roles.
- Promoting lifelong learning and continuous skill development to keep pace with technological advancements.

Policy and Governance:

- Developing regulatory frameworks that ensure the ethical and responsible use of AI technology.
- Encouraging public-private partnerships to drive innovation and address societal challenges.

Equitable Growth:

- Ensuring that the benefits of AI are widely distributed and accessible to all members of society.
- Addressing potential inequalities and ensuring that vulnerable populations are not left behind in the AI-driven future.

Practical Tips

- **Invest in Education:** Focus on acquiring skills relevant to an AI-driven future.
- **Advocate for Ethical AI:** Support initiatives that promote the ethical use of AI technology.
- **Embrace Lifelong Learning:** Stay adaptable and continuously seek new knowledge and skills.

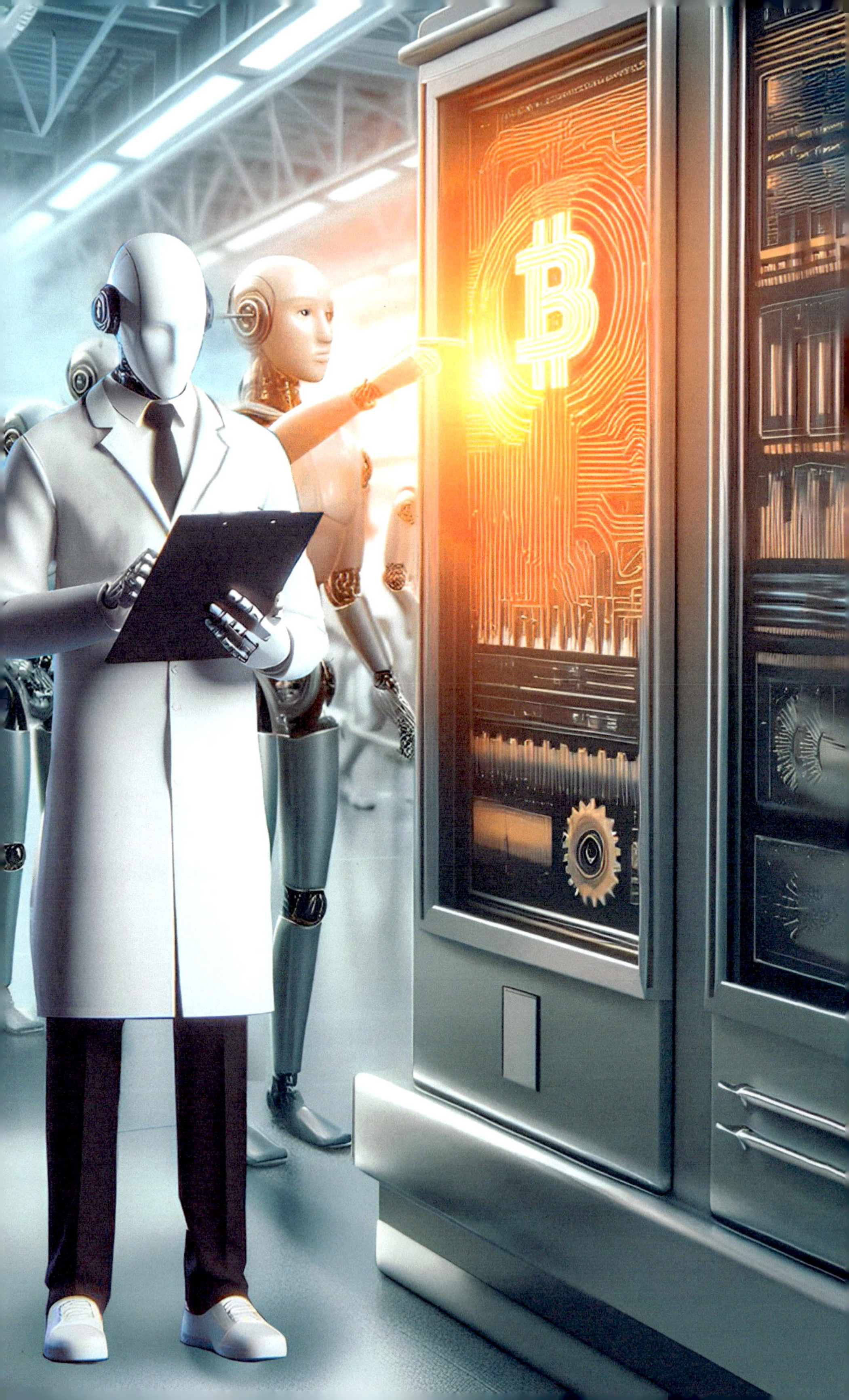

Chapter 11: What Will Happen to the Job Sector When Everyone Stays Home or Goes on Frequent Leisure Trips?

Shifts in Employment Patterns

The rise of remote work and increased leisure time facilitated by AI technology will lead to significant shifts in employment patterns and job sectors.

Remote Work and Flexibility:

- AI-driven tools and platforms enable remote work, allowing employees to work from anywhere with an internet connection.
- Increased flexibility may lead to a better work-life balance and more opportunities for leisure and personal pursuits.

- **Changes in Demand for Certain Jobs:**

- Industries that rely on in-person interactions, such as hospitality and retail, may face challenges as more people stay home.
- Demand for remote services, digital platforms, and online entertainment will increase, creating new job opportunities in these sectors.

Implications for the Workforce

The shift towards remote work and increased leisure time will have various implications for the workforce and the broader economy.

Workforce Demographics:

- Older workers who prefer traditional office environments may face challenges adapting to remote work.
- Younger workers and digital natives will likely thrive in remote, flexible work environments.

Economic Impact:

- Industries that cannot easily transition to remote work may experience labor shortages and increased costs to attract and retain employees.
- Companies may need to invest in technology and infrastructure to support remote work and ensure business continuity.

Policy and Organizational Responses

To navigate these changes, policymakers and organizations must develop strategies to support the evolving workforce and ensure economic stability.

Reskilling and Training Programs:

- Offering reskilling and training programs to help workers adapt to new job requirements and technologies.
- Encouraging continuous learning and professional development to keep pace with industry changes.

Support for Remote Work:

- Implementing policies and practices that support remote work, such as flexible scheduling, remote work allowances, and digital collaboration tools.
- Addressing potential challenges, such as cybersecurity risks and maintaining team cohesion in a remote work environment.

Practical Advice

- **Adapt to Remote Work:** Embrace remote work opportunities and develop skills to thrive in a virtual environment.
- **Seek Continuous Learning:** Engage in reskilling and professional development programs to stay competitive.
- **Support Workplace Flexibility:** Advocate for policies and practices that promote flexibility and work-life balance

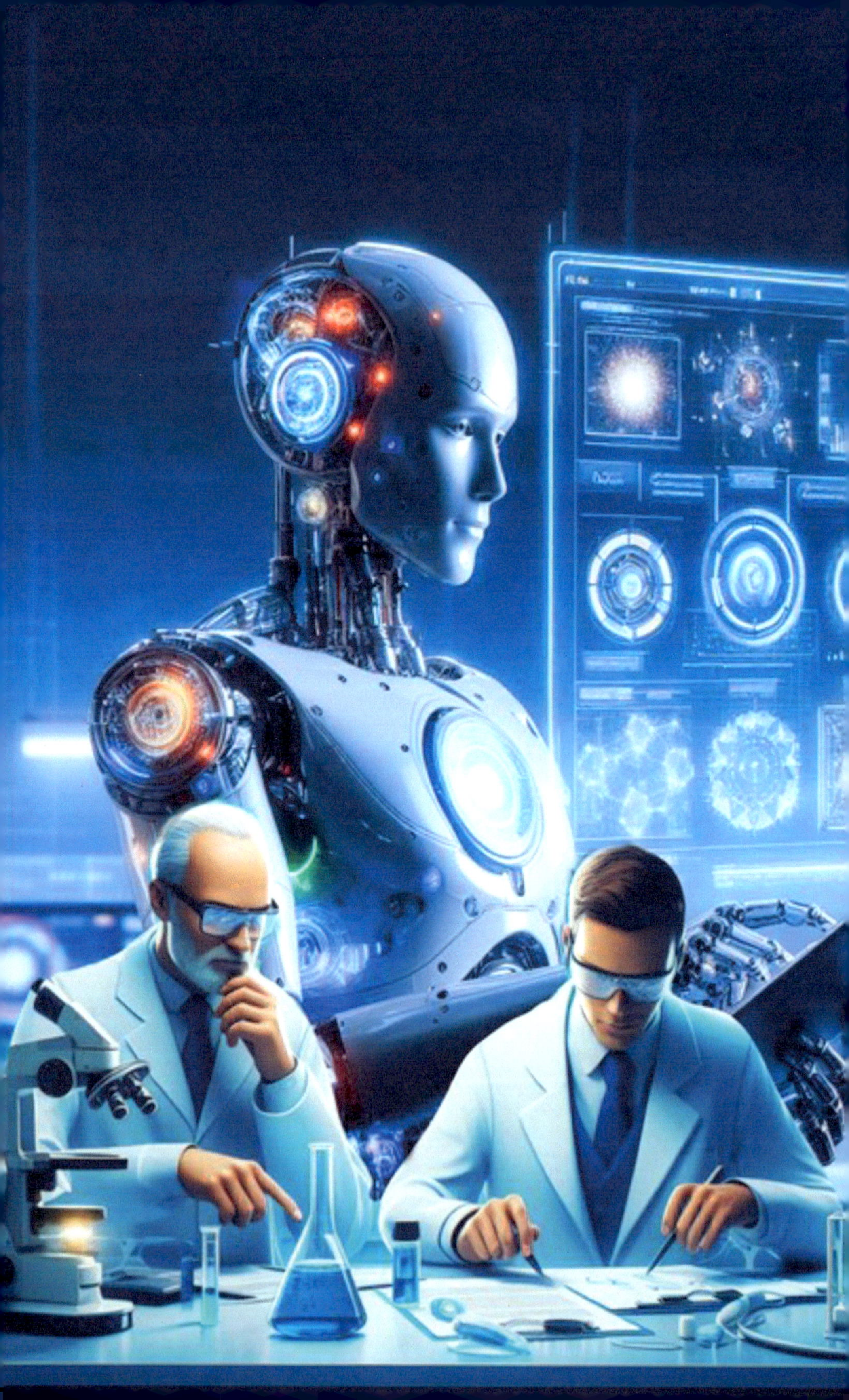

Chapter 12:
How Will AI Revolutionize the Healthcare Industry?

AI-Powered Healthcare Innovations

AI is set to revolutionize the healthcare industry by improving diagnostics, treatment, and patient care.

Enhanced Diagnostics:

- AI algorithms analyze medical images and data to accurately detect diseases at their early stages.
- Predictive models assess patient risk factors and health outcomes, enabling preventive care.

Personalized Treatment Plans:

- AI systems analyze patient data to create tailored treatment plans, optimizing therapy and medication.
- Precision medicine, driven by AI, ensures that treatments are effective and personalized to the patient's genetic makeup and health history.

Telemedicine and Remote Monitoring:

- AI-driven telemedicine platforms facilitate remote consultations, making healthcare accessible to those in remote or underserved areas.
- Wearable devices and AI-powered apps monitor patient health in real-time, alerting healthcare providers to potential issues.

Challenges and Considerations

While AI offers numerous benefits, there are challenges and ethical considerations to address in the healthcare sector.

Data Privacy and Security:

- Patient data privacy and security are paramount, given the sensitive nature of health information.
- Implementing robust cybersecurity measures to protect against data breaches and unauthorized access.

Bias and Fairness:

- AI systems must be trained on diverse datasets to avoid biases that could lead to unequal treatment outcomes.
- Continuous monitoring and validation of AI models are necessary to ensure fairness and accuracy.

Regulatory and Ethical Concerns:

- Developing regulatory frameworks to govern the use of AI in healthcare and ensure compliance with ethical standards.
- Engaging stakeholders, including patients, healthcare providers, and policymakers, in discussions about the ethical use of AI.

Practical Advice

- **Stay Informed:** Keep up with advancements in AI healthcare applications.
- **Advocate for Ethical AI:** Support the development of policies that ensure fair and ethical AI use in healthcare.
- **Embrace AI Tools:** Utilize AI-driven healthcare tools to improve personal health management and care.

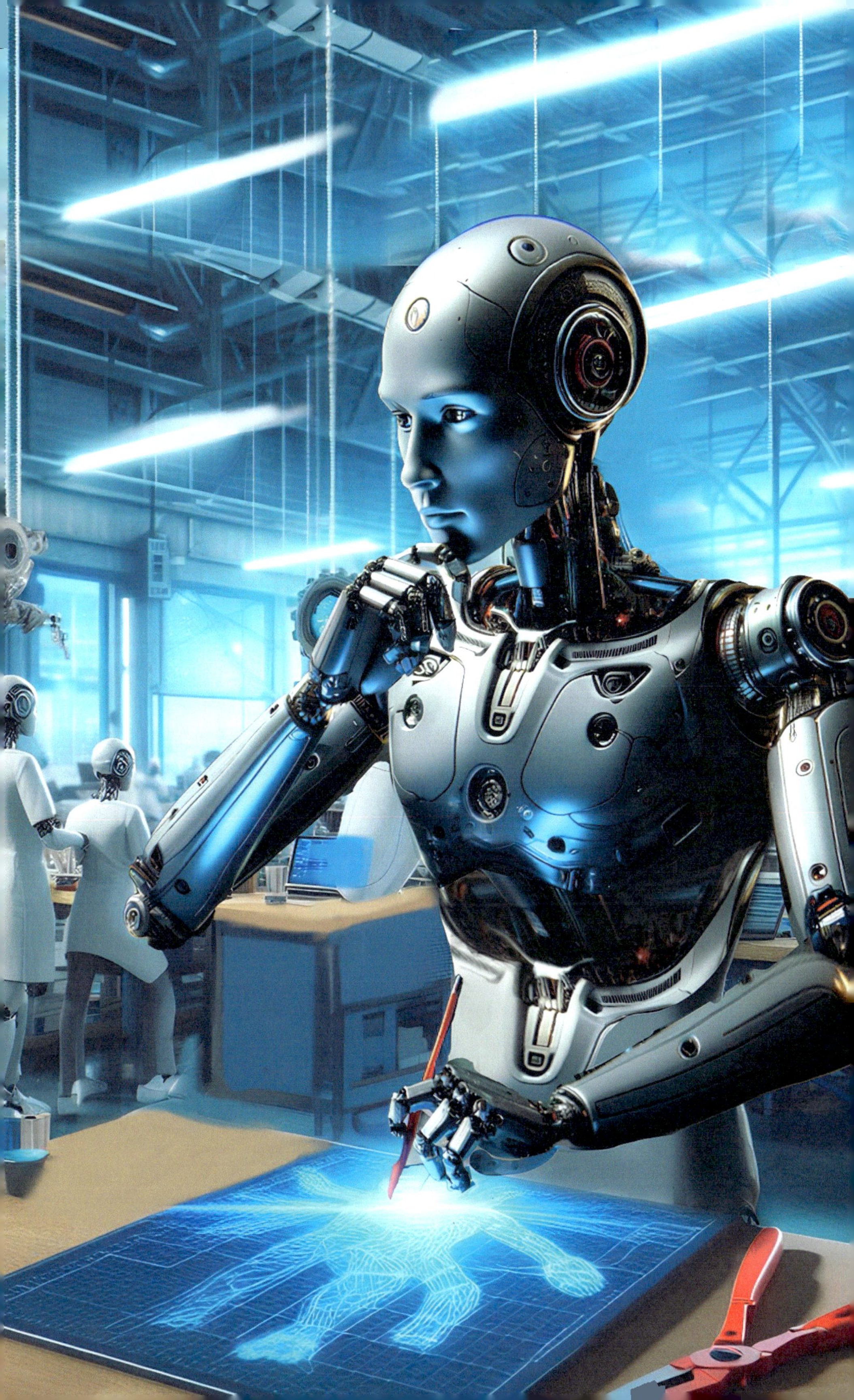

Chapter 13:
How Will AI Reshape Aviation and Traveling?

AI Innovations in Aviation

AI technology is transforming the aviation industry, enhancing safety, efficiency, and the passenger experience.

Autonomous and Smart Aircraft:

- AI systems enable autonomous flight operations, reducing the need for human pilots and minimizing the risk of human error.
- Smart aircraft equipped with AI can optimize flight paths, fuel consumption, and maintenance schedules.

Improved Air Traffic Management:

- AI algorithms analyze air traffic data to optimize flight routes and reduce congestion, leading to more efficient air travel.
- Predictive maintenance systems monitor aircraft health in real time, preventing delays and enhancing safety.

Transforming the Travel Experience

AI is also revolutionizing the travel experience, making it more personalized, convenient, and enjoyable.

Personalized Travel Services:

- AI-powered travel assistants provide personalized recommendations for flights, accommodations, and activities based on individual preferences.
- Real-time language translation tools facilitate communication in foreign countries, enhancing the travel experience.

Streamlined Travel Processes:

- AI-driven systems streamline airport check-in, security, and boarding processes, reducing wait times and enhancing passenger convenience.
- Contactless technologies and AI-powered health screening tools ensure safe and hygienic travel in the post-pandemic world.

Future Prospects

The future of AI in aviation and travel holds exciting possibilities, including:

Hyperloop and Advanced Transportation Systems:

- AI technology will be crucial in developing hyperloop systems and other advanced transportation modes, revolutionizing long-distance travel.

Global Connectivity and Accessibility:

- AI-driven innovations will enhance global connectivity, making travel more accessible and affordable for people worldwide.
- Efforts to reduce the environmental impact of travel through AI-optimized routes and sustainable practices will contribute to a greener future.

Practical Tips

- **Explore AI Tools:** Utilize AI-powered travel apps and services to enhance your travel experience.
- **Stay Informed:** Keep up with advancements in AI aviation technologies.
- **Advocate for Sustainability:** Support AI innovations that promote sustainable and eco-friendly travel.

Chapter 14: Banks and Independent Financial Institutions

AI in Financial Services

AI technology is transforming the financial services industry, enhancing efficiency, security, and customer experience.

Automated Financial Advisors and Chabot's:

- AI-powered financial advisors provide personalized investment advice and portfolio management, making financial planning accessible to a broader audience.
- Chabot's handle customer inquiries and transactions, improving response times and customer satisfaction.

Fraud Detection and Risk Management:

- AI algorithms analyze transaction patterns to detect fraudulent activities and prevent financial crimes.
- Risk management systems powered by AI assess market trends and identify potential risks, enabling proactive decision-making.

Impact on Traditional Banking

The integration of AI is reshaping traditional banking models, leading to greater efficiency and new service offerings.

Operational Efficiency:

- AI automates routine tasks such as data entry, document processing, and compliance checks, reducing operational costs and improving accuracy.
- Predictive analytics optimize resource allocation and enhance decision-making processes.

Customer Experience:

- Personalized banking services powered by AI tailor product offerings and communication to individual customer needs and preferences.
- AI-driven mobile apps and online platforms provide seamless and convenient banking experiences.

Future of Financial Institutions

The continued integration of AI and emerging technologies will define the future of banking and financial services.

Open Banking and Fintech Collaboration:

- AI will facilitate open banking initiatives, enabling customers to access and manage their financial data across multiple platforms.
- Collaboration between traditional banks and fintech startups will drive innovation and expand service offerings.

Sustainable Finance:

- AI can support sustainable finance initiatives by analyzing environmental, social, and governance (ESG) data to guide investment decisions.
- Financial institutions will be crucial in promoting sustainability through AI-driven solutions and responsible lending practices.

Practical Advice

- **Utilize AI Tools:** Explore AI-driven financial tools for personalized advice and efficient banking.
- **Stay Updated:** Keep abreast of AI advancements in the financial sector.
- **Advocate for Ethical AI:** Support the development of ethical AI practices in banking and finance.

Chapter 15: What Will Become of Suffering People Who Have No Access to AI and Little to No Access to Food and Water?

Addressing Inequality

As AI technology advances, it is crucial to address the widening gap between those with access to AI and those without access.

Impact on Vulnerable Populations:

- People in remote or impoverished areas may lack access to AI-driven services and opportunities, exacerbating existing inequalities.
- Limited access to education and technology can hinder the ability of vulnerable populations to benefit from AI advancements.

Strategies for Inclusion:

- **Digital Inclusion Programs:** Implementing programs that provide access to technology and digital literacy training in underserved communities.
- **Affordable Technology Solutions:** Developing affordable AI-powered solutions tailored to the needs of low-income populations.

Humanitarian Applications of AI

AI technology can be leveraged to address critical issues such as food security, water scarcity, and healthcare access.

AI for Food and Water Security:

- AI-driven agricultural technologies optimize crop yields, improve resource management, and enhance food production in challenging environments.
- Predictive models and IoT sensors monitor water resources, enabling efficient water management and distribution.

AI in Humanitarian Aid:

- AI-powered systems analyze data to identify and respond to humanitarian crises, ensuring timely and effective aid delivery.
- Drones and robotics assist in disaster response and relief efforts, reaching areas that are difficult to access.

Ethical and Social Considerations

Ensuring that AI benefits all members of society requires addressing ethical and social considerations.

Ethical Use of AI:

- Establishing guidelines for the ethical use of AI in humanitarian and development contexts.
- Engaging local communities in designing and implementing AI solutions to ensure cultural sensitivity and relevance.

Collaborative Efforts:

- Collaborating between governments, NGOs, and private sector partners to develop and deploy AI solutions for social good.
- Promoting international cooperation to address global challenges and ensure equitable access to AI technologies.

Practical Advice

- **Support Inclusion Programs:** Advocate for and support programs promoting digital inclusion and technology access.
- **Engage in Ethical AI Practices:** Encourage developing and using ethical AI solutions in humanitarian efforts.
- **Foster Collaboration:** Participate in initiatives that bring together diverse stakeholders to address global challenges with AI.

Chapter 16:
What Comes Next After AI?

Beyond AI: Emerging Technologies

As AI continues to evolve, new technologies will further transform our world.

Quantum Computing:

- Quantum computers have the potential to solve complex problems that are currently beyond the reach of classical computers.
- Advancements in quantum computing could revolutionize cryptography, materials science, and drug discovery.

Brain-Computer Interfaces (BCIs):

- BCIs enable direct communication between the brain and external devices, opening new possibilities for medical treatments and human augmentation.
- Applications of BCIs include restoring mobility to individuals with paralysis and enhancing cognitive abilities.

Advanced Robotics:

- Robotics technology will continue to advance, leading to more sophisticated and capable robots that can perform a wider range of tasks.
- Collaborative robots (cobots) will work alongside humans in various industries, enhancing productivity and safety.

Ethical and Societal Implications

Developing these emerging technologies will raise ethical and societal questions that must be addressed.

Ethical Considerations:

- Ensuring emerging technologies are developed and used ethically, focusing on human well-being and societal benefit.
- Addressing potential risks and unintended consequences, such as job displacement, privacy concerns, and security threats.

Societal Impact:

- Promoting inclusive and equitable access to emerging technologies to prevent further widening of the digital divide.
- Encouraging public dialogue and engagement to shape the development and deployment of new technologies in a way that aligns with societal values.

Preparing for the Future

It is essential to adopt a forward-thinking approach to navigate the rapidly changing technological landscape.

Investing in Research and Development:

- Supporting research and innovation in emerging technologies to drive progress and maintain a competitive edge.
- Encouraging interdisciplinary collaboration to address complex challenges and harness the full potential of new technologies.

Fostering Adaptability and Resilience:

- Cultivating a culture of adaptability and continuous learning to prepare individuals and organizations for future technological disruptions.
- Developing policies and frameworks that promote resilience and ensure that society can effectively respond to technological advancements.

Practical Tips

- **Stay Informed:** Keep up with developments in emerging technologies to understand their potential impact.
- **Engage in Ethical Discussions:** Participate in conversations about the ethical implications of new technologies.
- **Prepare for Change:** Develop skills and strategies to adapt to new technological landscapes.

Chapter 17: A Complete Guide to Understanding and Living with ChatGPT/AI

Understanding ChatGPT and AI

ChatGPT is a prime example of advanced AI technology that can assist with various tasks, from answering questions to generating creative content.

How ChatGPT Works:

- ChatGPT is based on the GPT (Generative Pre-Trained Transformer) architecture, which uses machine learning to generate human-like text.
- The model is trained on vast amounts of data and fine-tuned to understand and respond to various queries.

- **Applications of ChatGPT:**

- Personal Assistance: ChatGPT can help schedule, remind, and answer general knowledge questions.
- Content Creation: It can generate articles, stories, and other creative content based on user prompts.
- Learning and Education: ChatGPT can provide explanations, tutoring, and support for students and learners.

Integrating ChatGPT into Daily Life

To make the most of ChatGPT and similar AI tools, it is essential to understand how to integrate them into daily routines effectively.

Tips for Using ChatGPT:

- Clearly articulate your questions or prompts for the most accurate and relevant responses.
- Use ChatGPT to supplement human expertise, especially for complex or sensitive issues.
- Explore different applications and features to discover new ways AI can assist with tasks and projects.

Balancing AI Use

- **Set Boundaries:** Prevent over-reliance on AI by setting boundaries and ensuring it complements rather than replaces human interaction and decision-making.
- **Mindful Use:** Be aware of AI tools' limitations and potential biases and cross-check information with reliable sources when necessary.

Ethical Considerations and Best Practices

Using AI responsibly involves understanding the ethical implications and following best practices to ensure positive outcomes.

Ethical Use of AI:

- Respect privacy and confidentiality when using AI tools, especially when handling sensitive information.
- Be aware of potential biases in AI responses and cross-check information with reliable sources when necessary.

Continuous Learning:

- Stay informed about the latest developments in AI technology to make the most of new features and improvements.
- Engage in discussions about AI's ethical and societal impacts to contribute to responsible and informed AI use.

Practical Tips

- **Explore Applications:** Experiment with various AI tools to find the best fit for your needs.
- **Stay Updated:** Keep up with advancements in AI to leverage new features and improvements.
- **Engage Ethically:** Use AI responsibly, respecting privacy and being mindful of potential biases.

Made in the USA
Monee, IL
05 November 2026

67378698R00036